the Birds of Lord Howe Island

Ian Hutton

Introduction

Lord Howe Island is a haven for birdlife and this was one factor in recognising the Island for World Heritage Listing in 1982.

The forests provide food, shelter & breeding sites for 18 species of landbirds. The surrounding ocean provides abundant food each spring and summer for hundreds of thousands of seabirds from 14 species. In addition there are regular visitors easily seen for some months of the year, plus a number of irregular visitors that stop occasionally.

A total of 202 different birds has been recorded on the Island. This guide provides notes and sketches for all breeding species plus the regular visitors. In the text, the symbols ◯ and ¥ are used to indicate the length of the incubation and fledging periods, respectively. On the outer edge of each page is indicated the months during which the birds can be seen on the Island (or, in the case of the landbirds, the months during which they breed). Pages 52 to 54 give a checklist of all birds recorded on Lord Howe Island.

Walking tracks take you through the forests, and with a little patience many landbirds can be seen easily. Good seabird locations are: Ned's Beach for Flesh-footed shearwaters, Black-winged petrels and Sooty terns; North Bay for Sooty terns, Red-tailed tropicbirds, Black and Brown noddies; Malabar for Red-tailed tropicbirds; Little Island or Mt Gower for Providence petrels, Blackburn Island for Wedgetailed shearwaters; Muttonbird Point for Masked boobies, Sooty terns and Brown noddies. A round-island cruise passes many seabird colonies on remote headlands or offshore islets. Boat trips to Ball's Pyramid can be arranged, subject to weather.

Contents

Introduction 2
Seabirds 4
Landbirds 19
Map of Lord Howe Island 28-29
Regular non-breeding visitors 38
Checklist of birds on Lord Howe Island 52
Index 55

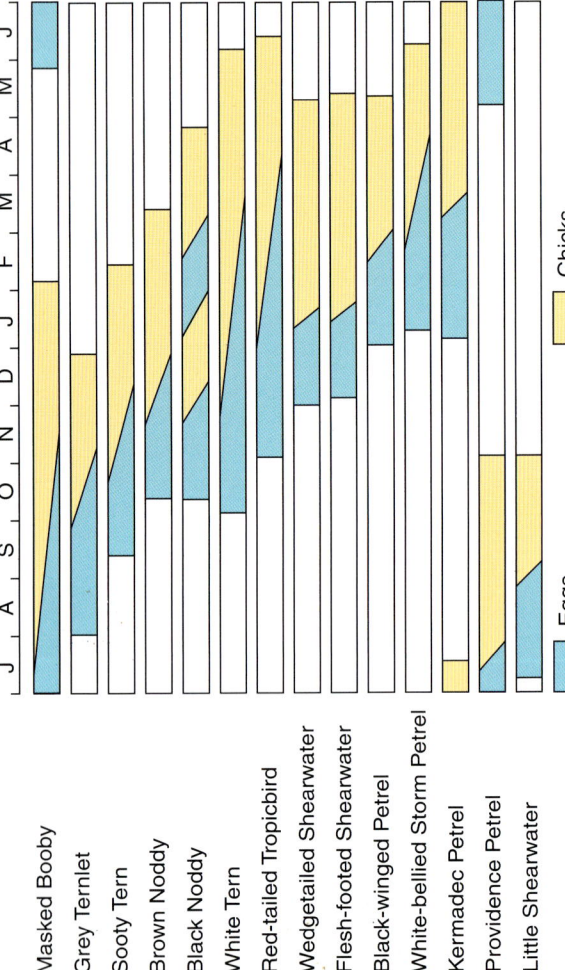

Seabirds

Providence Petrel   *Pterodroma solandri*

A dark-grey, winter breeding seabird, Lord Howe Island is the only place this bird breeds apart from a few pairs off Norfolk Island.

Characteristics: Length 40cm; wingspan 94cm; bill 35mm weight 500gm; dark grey upper, paler below, white scaly feathers around face, cream triangular patches under wings.
Habitat: Winter nesting around the southern mountains, flying around summits from midday, lower late in day.
Reproduction: Nest in burrow, 1 egg. ◐ 50 days ¥ 4 months.
Food: Fish, squid, crustaceans.
Flight: Slow wingbeats, long glides.

J F M A M J J A S O N D

Seabirds

Kermadec Petrel *Pterodroma neglecta*

A boat trip to Ball's Pyramid, 23km southeast of Lord Howe Island is usually necessary to see this bird, as it flies in and out of ledges high up on this remarkable rock stack.

Characteristics: Length 38cm; wingspan 92cm; bill 29mm; weight 500gm; dark brown all over with white flashes on top and under wings (also a light phase with white underbody).
Habitat: Breeds on Ball's Pyramid, seen flying over ocean around Lord Howe Island.
Reproduction: Surface nest on ledges, 1 egg 🥚 52 days ⚲ 15 weeks.
Food: Squid, crustaceans.
Flight: More ponderous than providence petrel.

Seabirds

Black-winged Petrel *Pterodroma nigripennis*

A small black and white petrel seen performing aerial courtship flights around its breeding colonies at Ned's Beach and Blinky Beach during spring and summer afternoons.

Characteristics: Length 30cm, wingspan 65cm, bill 24mm, weight 170gm. Upper body grey, under white, upper wing grey, underwing white with black margins, head black, throat and face white.
Habitat: Performs courtship flights around nests on cliffs Ned's Beach to Blinkie Beach, Mt Eliza, Erskine Valley.
Reproduction: Burrow in bushes on cliffs east coast. 1 egg.
🥚 45 days ⚥ 85 days
Food: Fish, crustaceans and squid.
Flight: Rapid and strong; wheeling courtship flight.

J F M A M J J A S O N D

J F M A M J J A S O N D

Seabirds

Flesh-footed Shearwater   *Ardenna carneipes*

The largest of the "muttonbirds" to breed at Lord Howe Island, it has its summer breeding colonies in the forest on the east coast from Ned's Beach to Clear Place.

Characteristics: Length 46cm; wingspan 102cm; bill 42mm; weight 600gm; entirely blackish brown; bill heavy and straw-coloured, feet flesh-coloured.
Habitat: Feeds at sea during day, returns at sunset, noisy all night at colonies in the forest from Ned's Beach to Clear Place.
Reproduction: In 2m long burrow, 1 egg. 🥚 57 days ¥ 90 days.
Food: Fish and squid.
Flight: Long glides close to surface, broken by effortless flaps.

Seabirds

Wedge-tailed Shearwater   *Ardenna pacifica*

A lighter shearwater, more buoyant in flight than the Flesh-footed, this species breeds mainly on the islands offshore from Lord Howe, but has been breeding along the lagoon dunes and Signal Point in recent years.

Characteristics: Length 43cm; wingspan 100cm; bill 40mm; weight 400gm; entirely sooty brown; tail wedge-shaped, bill slender, dark grey.
Habitat: Feeds at sea during day, returns to colonies from mid-afternoon, noisy at colony.

Reproduction: In burrow or grass tunnel, 1 egg.
🥚 53 days ¥ 90 days.
Food: Fish and squid.
Flight: Unhurried, slow flaps then long glide upwards, banking down.

J F M A M J J A S O N D

Seabirds

Little Shearwater *Puffinus assimilis*

A small shearwater, black above and white under, it has only re-colonised the main island at Lord Howe since cats were removed in the 1980s.

Characteristics: Length 28cm; wingspan 62cm; bill 23mm; weight 155gm. Crown and upper surfaces bluish black; underparts chalky white, underwings white, with black margins.
Habitat: Feeds at sea in groups, rarely seen ashore during the day; return to nests at night Ned's and Blinkie Beach cliffs.
Reproduction: In short burrow, 1 egg. 🥚 56 days ⚥ 72 days.
Food: Fish, squid, crustaceans.
Flight: Skims water with short glides and rapid wingbeats.

Seabirds

White-bellied Storm Petrel   *Fregetta grallaria*

The smallest of the seabirds to breed at Lord Howe Island, this bird is about the size of a canary. It breeds on the offshore islets and usually only seen when taking a boat trip to Ball's Pyramid.

Characteristics: Length 20cm; wingspan 40cm; bill 13mm; weight 48gm. Head, throat and upper parts black, belly and underwing white.
Habitat: Feeds well out to sea and only at nest on offshore islands at night.

Reproduction: Nest a cavity in rocks, 1 egg. 🥚 37 days ¥ 80 days.
Food: Crustaceans, squid.
Flight: Swift, close to sea; when feeding hangs in breeze with feet dangling, pattering the water.

J F M A M J J A S O N D

Seabirds

Masked Booby *Sula dactylatra tasmani*

The largest of the seabirds to breed at Lord Howe Island this seabird is not migratory and so can be seen at Lord Howe Island all year round.

Characteristics: Length 87cm; wingspan 170cm; bill 11cm; weight 2.2kg.
Habitat: Seen cruising past the Island, in and out of breeding sites at Muttonbird Point and offshore islands.

Reproduction: Flattened grass nest, 1 or 2 eggs. 🥚 43 days ⚲ 4 months.
Food: Fish and squid.
Flight: Long slow glides, slow wingbeats, dives down vertically from a height for food.

Seabirds

Red-tailed Tropicbird *Phaethon rubricauda*

A large, majestic white seabird with two long, stiff, red tail feathers. Seen flying off the cliffs during summer and autumn, Malabar to North Head and around the mountains and Ball's Pyramid.

Characteristics: Length 46cm (plus 2 tail feathers 50cm); wingspan 105 cm; bill 63mm; weight 700gm.
Habitat: Seen flying around breeding areas all day – cliffs Malabar to Mt Eliza, Nth Head, Mt Lidgbird and Mt Gower.

Reproduction: Nest is a scrape in soil on cliff ledge, 1 egg. 🥚 42 days ⚥ 10 weeks.
Food: Dive for fish and squid.
Flight: Ponderous and laboured flight, pairs perform aerial acrobatics in courtship flight.

J F M A M J J A S O N D

Seabirds

Sooty Tern   *Onychoprion fuscata*

A stark seabird, black above and white under, with a strongly forked tail. This is the most numerous of Lord Howe's seabirds and breeds on offshore islets and Ned's Beach to Middle Beach, Mt Eliza.

Characteristics: Length 46cm; wingspan 90cm; bill 42mm; weight 180gm; head and upper surfaces black, underparts white, tail deeply forked.
Habitat: When breeding is at colonies all day – offshore islets, Ned's Beach to Middle Beach, Mt Eliza, Muttonbird Pt, at sea feeding.
Reproduction: Scrape on bare ground, 1 egg. 🥚 28 days ⚥ 70 days.
Food: Snatch crustaceans, squid, fish from ocean surface.
Flight: Slow wingbeats; pairs perform parallel courtship flights.

Seabirds

Brown Noddy   *Anous stolidus*

The dark grey, slightly brown plumage is characteristic of this Noddy. It breeds on rocks and low bushes around the cliffs and offshore islands.

Characteristics: Length 40cm; wingspan 82cm; bill 42mm; weight 200gm, grey-brown body & wings, crown grey, forehead & eye-ring white.
Habitat: Seen at sea feeding in groups or at nesting colonies Old Gulch, Blinkie Beach, offshore islands; young birds 'loaf' on beaches late summer.
Reproduction: Twigs and seaweed on bushes or rock ledges, 1 egg. 🥚 35 days ✙ 50 days.
Food: Scoop crustaceans and fish from surface.
Flight: Swift with rapid wingbeats, low over water.

J F M A M J J A S O N D

Seabirds

Black Noddy *Anous minutus*

Smaller and blacker than the Brown Noddy, this species has a small breeding colony behind North Beach in the large Norfolk Island Pine trees.

Characteristics: Length 36cm; wingspan 70cm; bill 42mm; weight 100gm, sooty black body and wings, forehead and crown silver-white.
Habitat: Seen in groups or alone at sea; flying into colony at North Beach.

Reproduction: Substantial nest of seaweed, excreta, 1 egg. 🥚 35 days ⚲ 50 days.
Food: Snatch fish and crustaceans from surface.
Flight: Swift with rapid wingbeats low over water.

Seabirds

White Tern *Gygis alba*

Often described as angelic, this pretty white seabird is a favourite with residents and visitors alike. Its habit of breeding low down on trees along roadsides makes it very easy to observe closely.

Characteristics: Length 30cm; wingspan 75cm; bill 40mm; weight 110gm. White all over, thin black circle around eyes, forked tail.
Habitat: Colonies in the settlement area particularly pine trees along the lagoon; North Bay.

Reproduction: Balances 1 egg on branch. ◗ 28 days ⚥ 70 days.
Food: Plucks small fish and squid from the surface, can hold up to 5 held crossways in bill.
Flight: Swift, rapid wingbeats, parallel courtship flights.

J F M A M J J A S O N D

Seabirds

Grey Ternlet   *Procelsterna cerulea*

A small light grey ternlet, providing an unforgettable experience when you see groups of them sitting low down on the black basalt cliffs around the Island late in the day.

Characteristics: Length 28cm; wingspan 52cm; bill 26mm; weight 70gm, body and wings light blue-grey, breast paler, primary feathers darker.
Habitat: Seen feeding at sea in flocks; sitting on cliffs around remote parts of Island.

Reproduction: Nest a few strands of seaweed on rock ledge; 1 egg. ◐ 32 days ϒ 37 days.
Food: Pluck small fish and crustaceans from surface.
Flight: Graceful and buoyant.

Landbirds

White faced Heron *Ardea novaehollandiae*

A tall, wary, grey bird with a long neck and yellow legs; often seen foraging alone along the rocky seashore or remote paddocks of the Island.

Characteristics: Length 67cm; sexes similar, body blue-grey, wings darker, face white, legs and feet yellow-olive, bill black.
Habitat: swampy areas, remote paddocks, rocky seashore at low tide.

Reproduction: Nest a large platform of sticks hidden in tree tops; 3-5 eggs. 🥚 25 days ❌ 40 days.
Food: Worms, spiders, snails, fish, crustaceans.
Flight: Low with slow, laboured wingbeats.

J F M A M J J A S O N D

Landbirds

Australian Kestrel *Falco cenchroides*

A small, brown bird of prey seen hovering motionless 10 to 20 m above the ground, into the wind as it seeks out food over remote pastures and cliffs.

Characteristics: Length: male 31cm, female 35cm; upperparts brown, face white, underparts white, tail grey, bill blue-grey, feet yellow.
Habitat: Seen over remote paddocks and cliffs.

Reproduction: Scrape on cliff ledge or in tree hollow, 3-4 eggs. 🥚 26 days ♀ 26 days.
Food: Insects, mice, seabird chicks.
Flight: Hover motionless into wind, descending by stages onto prey.

Landbirds

Woodhen *Tricholimnas sylvestris*

A large, brown flightless bird with a long drooping bill. Once one of the rarest birds in the world, now widespread over Lord Howe Island thanks to a captive breeding program in the early 1980s.

Characteristics: Length 36cm, body olive-brown, flight feathers barred chestnut and black.
Habitat: All over Island where palms and water available, settlement area.
Reproduction: Nest of grass, moss, palm fibre under tree roots or Providence petrel burrows, 1-4 eggs. 🥚 20 days.
Food: Worms, molluscs, invertebrates in leaf litter.
Flight: Flightless although may use wings to 'hop' up logs and small ledges.

J F M A M J J A S O N D

J F M A M J J A S O N D

Landbirds

Buff-banded Rail   *Gallirallus philippensis*

A small colourful rail, slightly smaller than the Woodhen and has flecks of white and black on feathers; becoming more common and less shy around the settlement.

Characteristics: Length 30cm; upperparts olive mottled with black and white; crown olive-brown, face grey, chestnut band through eye, throat grey; under barred black and white with chestnut band across breast.

Habitat: Shy, common through settled lowland areas.
Reproduction: Cup-shaped grass nest, 5-8 eggs. 🥚 20 days.
Food: Insects, molluscs and plants.
Flight: May fly short distances with legs dangling.

Landbirds

Masked Lapwing *Vanellus miles*

This bird is the most recent coloniser to Lord Howe — only in 1991 was it first breeding. It feeds in open pasture and is noticeable with its erect stance and staccato call.

Characteristics: Length 36cm; upperparts brown, white under, black crown extending down neck to sides of back. Face white with yellow wattles, a buff, spur on each shoulder.

Habitat: Grassy pasture, airstrip, golf course.
Reproduction: Nest a depression in grass, 3-4 eggs. 28 days.
Food: Insects, worms, beetles, spiders.
Flight: Slow, deliberate wingbeats, calls in flight.

Landbirds

Purple Swamphen *Porphyrio porphyrio*

A large, brightly coloured black and blue bird with a large red bill; seen along the pasture edge at the southern parts of the Island. It has only colonised Lord Howe in 1987.

Characteristics: Length 48cm; upperparts and face black; neck, breast and belly deep blue, undertail white, large bill and frontal shield red.
Habitat: Edge of pasture around golf course and airstrip.

Reproduction: Nest of pulled-down grass, 3-8 eggs. 26 days.
Food: Plant roots, insects, molluscs, grubs.
Flight: Laboured low flight with legs dangling.

Landbirds

Masked Owl   *Tyto novaehollandiae*

Introduced in the 1920s to control rats, this medium sized owl is rarely seen because of its nocturnal habits, although occasionally heard at night on the forest edge.

Characteristics: Length, male 40cm, female 50cm; upperparts blackish brown with speckles of white, underparts pale buff to white; face disc cream to buff.
Habitat: Forest away from settlement, nocturnal.

Reproduction: Nest in tree hollows, 2-3 eggs. 🥚 35 days.
Food: Mice, rats, birds.
Flight: Swift, silent flight.

J F M A M J J A S O N D

Landbirds

Emerald Ground-dove *Chalcophaps indica*

A ground-feeding bird with bright green upper wings, very tame and often seen wandering nonchalantly about the forest and roadsides.

Characteristics: Length 26cm; back and underparts brown, wings and shoulders bright green with white elbow patches.
Habitat: Solitary, mainly lowland forest, feeding on the ground, tame around settlement.

Reproduction: Nest a low platform of twigs and vines, 2 eggs. ⬭ 15 days.
Food: Fallen fruits.
Flight: Low, fast, rapid wingbeats.

Landbirds

Welcome Swallow *Hirundo neoxena*

A small, fast-flying bird seen swooping low over pasture to catch insects on the wing; also sitting in a row on fence wires. Only colonised LHI in the early 1970s.

Characteristics: Length 15cm, Shiny blue-black upperparts, dull black wings, throat and forehead rufous, underparts fawn.
Habitat: Flying low over pasture and swamps.

Reproduction: Mud nest in buildings or coastal caves, 4-5 eggs. 🥚 15 days ✱ 20 days.
Food: Catches insects on the wing.
Flight: Fast, low, swooping flight.

J F M A M J J A S O N D

Admiralty Islets
Old Gulch
KIMS LOOKOUT
Mt ELIZA 147m
MALABAR 209m
Neds Beach
North Bay
North Head
PO
Middle Beach
REEF
LAGOON
Clear Place
TRANSIT HILL 121m
Blackburn Island
Blinky Beach
Mutton Bird Is

LORD HOWE ISLAND

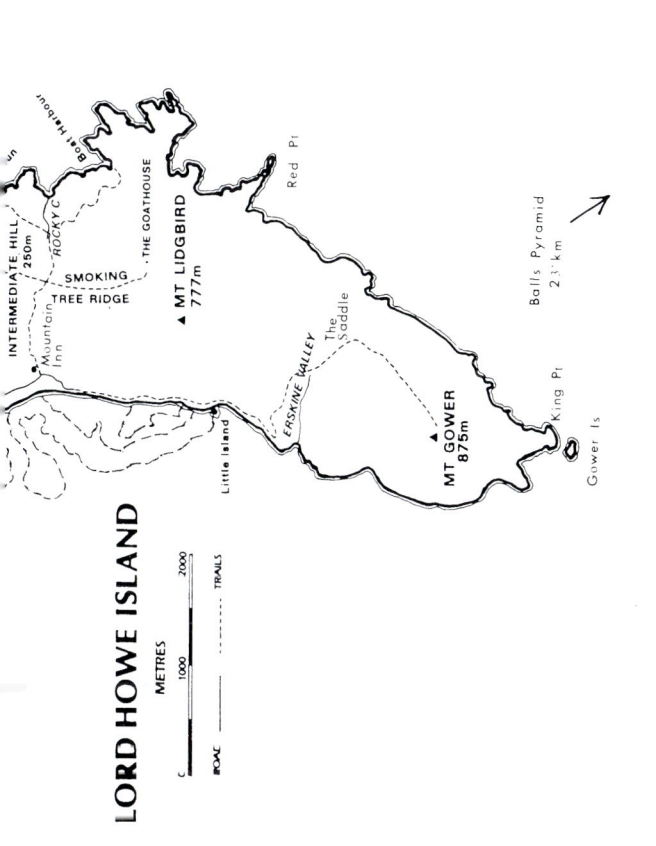

Landbirds

Sacred Kingfisher   *Todiramphus sanctus*

The striking blue/green back of this bird is unmistakable when seen in flight or from behind. They are often observed sitting on fence posts along the airstrip.

Characteristics: Length 20cm; head, back and wings blue or green-blue, black band bill to nape of neck; underparts and collar buff, throat white.
Habitat: Sits on fence posts eyeing off food, also on rocks of seashore.

Reproduction: Nest in tree hollow, 3-6 eggs. ◐ 17 days ¥ 24 days.
Food: Worms, insects, fish, crustaceans, mice, snails, small birds.
Flight: Low, rapid wingbeats and glides.

Landbirds

Song Thrush   *Turdus philomelos*

A native of Europe, introduced to Australia and New Zealand in the 1850s, and introduced to Lord Howe Island 1944. Not common on the Island, confined to the settlement area this bird also adds its melody to the dawn chorus.

Characteristics: Length 23cm, upperparts olive-brown throat pale yellow, underparts white with small brown triangles.
Habitat: Lowland forest around settlement.

Reproduction: Nest cup of grass, palm fibre, leaves lined with mud, 4-5 eggs. 14 days ¥ 15days.
Food: Ground feeders on fruits, seeds, insects, spiders, beetles, moths.
Flight: Low, rapid wingbeats.

Landbirds

Blackbird   *Turdus merula*

A small, black bird, common all over the settlement area, with a beautiful call dawn and dusk. Native to Europe, introduced to Australia and New Zealand in the 1850s, and introduced to Lord Howe Island 1944.

Characteristics: Length 25cm. Male black all over, yellow eye ring. Female dark brown, grey chin.
Habitat: Lowland forest around settlement area.

Reproduction: Nest of grass in trees, shrubs; 3-4 eggs. 🥚 14 days ❤ 14 days.
Food: Ground feeders on insects, worms, fruits, berries.
Flight: Low with rapid wingbeats.

Landbirds

Lord Howe White-eye — *Zosterops tephropleura*

A small but abundant bird in the lowlands and mountains, often seen in groups flitting from tree to tree in search of insects and nectar.

Characteristics: Length 13cm, head and throat yellow, thin white eye ring, back grey, belly whitish, upper wings and rump green-yellow.
Habitat: Concealed in foliage of forest trees all over island.

Reproduction: Small cup nest of palm fibre and spider web, 3 eggs. ◐ 12 days ⚥ 12 days.
Food: Insects, nectar, fruits, seeds.
Flight: Rapid flight through foliage.

J F M A M J J A S O N D

Landbirds

Lord Howe Golden Whistler   *Pachycephala pectoralis contempta*

The most common native bird seen and heard while on the lowland forest walks, the male has a bright yellow chest with black upper band and is quite striking.

Characteristics: Length 17cm. Male head and face black, broad yellow collar, throat white with black band below. Back and wings olive-grey, belly yellow. Female upperparts olive-grey, underparts grey with yellowish wash.

Habitat: Forest all over the Island, tame around settlement.
Reproduction: Cup nest of palm fibre and leaves, 2 eggs. 15 days ⚥ 13 days.
Food: Hop branch to branch feeding on insects and larvae.
Flight: Fast, low, rapid wingbeats interspersed with short glides.

Landbirds

Magpie Lark (Peewee) *Grallina cyanoleuca*

A familiar white and black bird seen walking around the pasture areas in search of food. Introduced from Australia 1920s.

Characteristics: Length 28cm, Male upperparts black with white patch on wing. Rump and underparts white, black head throat and upper breast. Female same but forehead and throat white.

Habitat: Grassy paddock areas and forest edge.
Reproduction: Mud nest high on branch, 3 eggs. ⬬ 18 days ⚲ 20days.
Food: Ground feeders on insects and larvae.
Flight: Slow at height.

J F M A M J J A S O N D

Landbirds

Common Starling *Sturnus vulgaris*

A stocky, dark bird with a short tail, upright stance and a jaunty walk. Only occurs on Lord Howe Island in small numbers mainly seen at Old Settlement paddock.

Characteristics: Length 21cm; black all over with a green-purple sheen except for a brown wash on wings and tail. New winter plumage has buff speckles on body. Bill black, yellow in breeding; feet brown, eyes brown.

Habitat: Old Settlement and Golf Course.
Reproduction: Untidy grass nest in tree hollow or palm crown; 4-7 eggs. ⬤ 12 days ⚥ 21 days.
Food: Insects, fruits, seeds.
Flight: Fast with rapid wingbeats.

Landbirds

Lord Howe Island Currawong *Strepera graculina crissalis*

A large black bird of the forest, inquisitive and often follows walkers along trails, watching with beady eyed curiosity. This subspecies is unique to Lord Howe Island.

Characteristics: Length 46cm, sooty black with small white patches on wings and tail. This Lord Howe Island subspecies has a larger bill than mainland species.
Habitat: Forests all over the Island, particularly the southern mountain areas.

Reproduction: Bulky nest of sticks high in tree, 3 eggs.
◐ 21 days ⚲ 20 days.
Food: Native fruits and seeds, insects, grubs, chicks of other birds, lizards, snails.
Flight: High, slow wingbeats, often calling in flight.

J F M A M J J A S O N D

Visitors

Black-browed albatross   *Diomedea melanophrys*

The most common albatross seen off S.E. Australia. The Campbell Island race (with yellow eye) is seen offshore from Lord Howe Island on boat cruises autumn and winter.

Characteristics: Length 90cm; wingspan 230cm; bill yellow-orange with pink tip, 11cm; weight 3kg. Adult: head, neck, rump, underparts white; thin black eyebrow. Mantle, upperwing, tail black; underwing white with broad black leading edge, tip and trailing edge black. Juvenile: as adult but underwing mostly black; grey collar and crown; bill black-brown.
Habitat: Breeds on subantarctic islands and seen around LHI autumn and winter.
Reproduction: coned shaped nest of earth and vegetation; 1 egg, incubation 71 days; fledging 120 days.
Food: squid, fish, crustaceans, offal from boats.

Visitors

Wandering Albatross *Diomedea exulans*

The largest species of albatross is seen offshore from Lord Howe Island on boat cruises autumn and winter. The large size and slow soaring glides unmistakable.

Characteristics: Length 120cm; wingspan 340cm; bill pink, tipped yellow, 17cm; weight 8kg. Shows 7 colour phases from juvenile all chocolate brown except for white face and underwing, through to adult white above and below; breast has varying amount of fine black speckles. Upperwing white with black tip and trailing edge; underwing white with black tip; tail white.
Habitat: Breeds on subantarctic islands and seen around LHI autumn and winter.
Reproduction: cone-shaped nest of earth and vegetation; 1 egg, incubation 70-80 days, fledging 270 days.
Food: fish and squid, offal from boats.

J F M A M J J A S O N D

J F M A M J J A S O N D

Visitors

Pacific Golden Plover   *Pluvialis fulva*

Short, quick steps, interspersed with pauses of upright stance are characteristics of this summer visitor, which has mottled brown and golden plumage.

Characteristics: Length 24cm; wingspan 65cm; bill 22mm. Upperparts mottled brown and buff with golden flecks, forehead, face neck buff, underparts buff white. In breeding plumage face, throat, breast black, back stronger colours.

Habitat: Foraging on pasture or seashore, alone or small groups.
Breeds: Alaska, Siberia, northern Canada.
Food: Insects, molluscs, crustaceans, plant matter.

Visitors

Cattle Egret *Ardea ibis*

Unmistakable with its large size and white colour, this visitor is hard to miss as it struts around pasture areas following cattle.

Characteristics: Length 50cm. Non-breeding: white all over; bill pale yellow, feet olive-yellow, face skin green-yellow. In breeding crown, neck and back tinged orange-buff, bill red with yellow tip, face skin red.

Habitat: Paddock areas with grazing cattle.
Breeds: Northern New South Wales, crosses to New Zealand for winter, some stop briefly at Lord Howe Island.
Food: Insects and worms.

J F M A M J J A S O N D

Visitors

Ruddy Turnstone   *Arenaria interpres*

A small squat bird with upperparts mottled brown and black, white under. It is the most common of summer visitors to Lord Howe Island — seen at the airport turning over bits of mown grass in search of food.

Characteristics: Length 23cm, wingspan, 52cm, bill 22cm. Upperparts brown mottled with grey and black; throat and abdomen white, upper breast blackish; legs orange, bill black. Breeding plumage upperparts chestnut and black, head white.

Habitat: Open low grassy areas, sandy or rocky seashores low tide in groups 6-20.
Breeds: Arctic Circle, Alaska, nthn Canada, Greenland, Scandinavia, Russia.
Food: Insects, worms, crustaceans, molluscs.

Visitors

Double-banded Plover *Charadrius bicinctus*

A winter resident, this is the smallest of the visitors to Lord Howe Island and can be observed feeding along the airstrip verge or at low tide at North Bay.

Characteristics: Length 18cm, bill 18mm. Upperparts grey-brown; forehead, chin, throat, eyebrows and underparts white, remnant breeding bands at sides of breast. Breeding: 2 chest bands; upper black, lower chestnut. Legs grey-green, bill black.

Habitat: Open low grassy areas, sandy seashore low tide, alone or twos, threes.
Breeds: New Zealand, crosses Tasman Sea to Australia for winter.
Food: Insects, worms, crustaceans.

J F M A M J J A S O N D

Visitors

Whimbrel   *Numenius phaeopus*

This large summer resident, with variegated light and dark brown plumage, has excellent camouflage and is often heard before seen, as it calls out when flying off.

Characteristics: Length 41cm; wingspan 80cm; bill 9cm down-curved. Body buff with upperparts mottled darker grey-brown, throat and breast streaked with brown. Bill brown, down-curved; legs grey.

Habitat: Remote grassy paddocks, rocky seashores, alone.
Breeds: Above Arctic Circle.
Food: Insects, seeds, molluscs, worms, crustaceans.

Visitors

Eastern Curlew *Numenius madagascariensis*

A very large wader, with a disproportionately long, down-curved bill. Usually seen alone and may be seen at any time of the year, some may stay for several weeks.

Characteristics: Length 60cm, including bill 18cm long, black with pink base. Body streaked dark brown and buff above, slightly paler below. Feet and legs olive-grey.
Habitat: Seen feeding on the beach, rocky seashore, swampy areas or short grassy areas such as the golf course or airstrip verge.
Breeds: Siberia May and June; nest a scrape on a small mound, lined with grass.
Food: Worms, crabs and insects.

J F M A M J J A S O N D

Visitors

Wandering Tattler   *Tringa incarna*

A solitary wader, difficult to distinguish from Grey-tailed Tattler. The call may be the best guide — the Wandering has 6–10 notes on flying off, the Grey-tailed has just a double whistle.

Characteristics: Length 27 cm; bill dark grey, straight, 39 mm long; legs yellow. Upper body dark grey, breast grey, white below; white eyebrow doesn't extend far behind the eye.
Habitat: Mainly seen on the rocky shore of the main island & offshore islets, on seaweeds of the surf zones; usually alone, occasionally 2 or 3 together.
Breeds: NE Siberia & Alaska from May to August, migrating in winter to Pacific coast of America, Pacific islands, N.Z. and coastal E. Australia.
Food: Crustaceans, worms, molluscs, insects.

Visitors

Grey-tailed Tattler — *Tringa brevipes*

Usually a solitary wader, sometimes in group of 2 or 3; seen at Lord Howe Island October to April. Difficult to distinguish from Wandering Tattler. The call may be the best guide — the Grey-tailed has just a double whistle on flying off, the Wandering has 6–10 notes.

Characteristics: Length 23 cm; bill dark grey, straight, 38 mm long; legs yellow. Upper body light grey, breast grey, white below; a white eyebrow extending well beyond the eye.
Habitat: Mainly seen feeding at low tide on sea grass beds at North Bay; usually alone, occasionally 2 or 3 together.
Breeds: Remote mountains of eastern Siberia, migrating in winter to coastal southeast Asia, New Guinea and Australia.
Food: Crustaceans, worms, molluscs, insects.

J F M A M J J A S O N D

Visitors

Bar-tailed Godwit   *Limosa lapponica*

A large mottled grey and brown wader with a long straight bill which it uses to probe the sandy seashore or swampy paddocks for food.

Characteristics: Length 37-40cm; wingspan 70-80cm; bill 8cm, straight. Non-breeding: head and neck brown-grey; upperparts mottled brown and grey; breast buff-grey, underparts off white. Bill pink, grey tip; legs dark grey. Breeding males reddish head, neck and underparts, females tawny buff.
Habitat: Grassy areas or sandy seashore at low tide, in small groups of 4-5.
Breeds: Scandinavia to Northern Asia to Alaska.
Food: Worms, crustaceans, larvae, insects; probes with long bill.

Visitors

Japanese or Latham's Snipe *Gallingo hardwickii*

A typical snipe with a long bill and short legs, this shy summer visitor is wary and well camouflaged and so hard to find. Early or late in the day is the best time to search. Usually in a group of 3 to 5 on Lord Howe Island.

Characteristics: Length 25 cm; bill olive-brown, straight 68 mm; short olive-grey legs. Body mottled black, brown and buff; belly white; flanks barred, chestnut band on tail. Face pale cream with dark eye stripe and cheek stripe.
Habitat: Wet grassland areas e.g. Moseley Park swamp, creeks at Old Settlement, Cobbys Corner, Soldiers Creek.
Breeds: Japan from May to July and migrates south to eastern Australia. (On LHI August to April)
Food: Worms, insect larvae and seeds.

J F M A M J J A S O N D

J F M A M J J A S O N D

Visitors

Red-necked Stint *Calidris ruficollis*

The smallest of the waders to visit Lord Howe Island, this small summer visitor is easy to overlook. The most numerous of the arctic waders to visit mainland Australia, but only a few visit the Island.

Characteristics: Length 15 cm, wingspan 36 cm, bill olive-grey, 17 mm, short black legs. Back and wings mottled grey, head and side of neck shaded grey, underparts white. In March may start to show breeding plumage with red head and brown upperparts.

Habitat: Most often seen at low tide on the weed flats at North Bay, or at Moseley Park swamp.
Breeds: Northeast Siberia, June to August.
Food: Worms, molluscs, crustaceans and some seeds.

Visitors

Great Cormorant *Phalacrocorax carbo*

The largest of Australian Cormorants. Black all over with a yellow face. Not as common at Lord Howe Island as the Little Black Cormorant, but usually one or two seen most summers, often sitting with Little Black Cormorants.

Characteristics: Length 80cm. Black all over with yellow facial skin and throat pouch. Bill grey, buff at base; legs and feet black.
Habitat: At LHI seen sitting on rocks sunning at Far Rocks, mouth of Soldiers Creek, Middle Beach; sometimes in Moseley Park swamp.
Breeds: Australia and NZ in large nest of sticks and debris in a tree or on the ground.
Food: Dives for fish and crustaceans.

J F M A M J J A S O N D

Checklist

PROCELLARIIFORMES
Wandering Albatross ....R
Black-browed Albatross V
Bullers Albatross V
Indian Yellow-nosed Albatross .V
Northern Royal Albatross V
Shy Albatross V
Northern Giant Petrel ....IR
Southern Giant Petrel ....IR
Southern Fulmar V
Cape Petrel ....IR
Great-winged Petrel ....IR
White-headed Petrel V
Grey-faced Petel V
Providence Petrel ....B
Kermadec Petrel ....B
Tahiti Petel V
Mottled Petrel V
White-necked Petrel V
Black-winged Petrel ....B
Pycroft's Petrel X
Gould's Petrel V
Antarctic Prion V
Fairy Prion V
Westland Petrel V
Wedge-tailed Shearwater ......B
Buller's Shearwater V
Flesh-footed Shearwater ......B
Sooty Shearwater V
Short-tailed Shearwater V
Fluttering Shearwater V
Hutton's Shearwater V
Little Shearwater ....B
Wilson's Storm-Petrel V
White-faced Storm-Petrel .....?
White-bellied Storm-Petrel ...B

SPHENISCIFORMES
Little Penguin V

PODICIPEDIFORMES
Australasian Grebe V
Hoary-headed Grebe V

PELECANIFORMES
Red-tailed Tropicbird ....B
Great Frigate Bird V
White-tailed Tropicbird V
Australasian Gannet ....IR
Masked Booby ....B
Red-footed Booby V
Brown Booby V
Little Pied Cormorant V
Pied Cormorant V
Little Black Cormorant V
Great Cormorant ....R
Lesser Frigatebird V

CICONIIFORMES
White-faced Heron ....B
Little Egret V
Eastern Reef Egret V
Great Egret V
Intermediate Egret V
Cattle Egret ....R
Nankeen Night Heron V
Little Bittern V
Australasian Bittern V
Glossy Ibis V
Australian White Ibis V
Straw-necked Ibis V
Royal Spoonbill V
Yellow-billed Spoonbill V

FALCONIFORMES
Black-shouldered Kite V
Brahminy Kite V
Swamp Harrier V
Brown Falcon V
Nankeen Kestrel ....B

ANSERIFORMES
Black Swan V
Canada Goose V
Australian Shelduck V
Paradise Shelduck V
Australian Wood Duck V
Hardhead V
Mallard V
Pacific Black Duck V
Grey Teal V
Chestnut Teal V

GALLIFORMES
California Quail IX
Feral Chicken I

GRUIFORMES
Buff-banded Rail ....B
Woodhen ....B
Baillon's Crake V
Purple Swamphen ....B
White Gallinule X
Dusky Moorhen V
Eurasian Coot V

CHARADRIIFORMES
Latham's Snipe ....R
Black-tailed Godwit V
Bar-tailed Godwit ....R
Little Curlew V
Whimbrel ....R
Eastern Curlew ....R
Marsh Sandpiper V
Common Greenshank V
Terek Sandpiper V
Common Sandpiper V
Grey-tailed Tattler ....R
Wandering Tattler ....R
Ruddy Turnstone ....R
Great Knot V
Red Knot ....IR
Red-necked Stint ....R
Pectoral Sandpiper V
Sharp-tailed Sandpiper ......IR
Curlew Sandpiper V
Sanderling V
Buff-breasted Sandpiper V
Painted Snipe V
South Island Pied Oystercatcher ..V
Sooty Oystercatcher V
Black-winged Stilt V
Pacific Golden Plover ....R
Grey Plover V
Double-banded Plover ....R
Lesser Sand Plover V
Greater Sand Plover V
Oriental Plover V
Black-fronted Dotterel V
Banded Lapwing V
Masked Lapwing ....B
Oriental Pratincole V
Australian Pratincole V
Great Skua V
Long-tailed Jaeger V
Kelp Gull V
Silver Gull V
Laughing Gull V
Gull-billed Tern V
Caspian Tern V
Crested Tern V
White-fronted Tern V
Black-naped Tern V
Common Tern V

Arctic Tern V
Roseate Tern V
Little Tern V
Sooty Tern ....B
Whiskered Tern V
White-winged Black Tern V
Common Noddy ....B
Black Noddy ....B
Grey Ternlet ....B
White Tern ....B

COLUMBIFORMES
Rock Dove ....B
White-throated Pigeon X
Spotted Turtle-dove V
Brush Bronzewing V
Emerald Dove ....B
Pied Imperial-Pigeon V

PSITTACIFORMES
Swift Parrot V
Eastern Rosella V
Red-crowned Parakeet X

CUCULIFORMES
Oriental Cuckoo V
Pallid Cuckoo V
Brush Cuckoo V
Fan-tailed Cuckoo V
Shining Bronze-Cuckoo V
Common Koel V
Long-tailed Cuckoo V
Channel-billed Cuckoo V

STRIGIFORMES
Barn Owl I
Masked Owl I
Southern Boobook I
LHI Boobook X

APODIFORMES
White-throated Needletail V
Fork-tailed Swift V

CORACIIFORMES
Sacred Kingfisher ....B
Rainbow Bee-eater V
Dollarbird V

PASSERIFORMES
LHI Gerygone X
Noisy Friarbird V
LHI Golden Whistler ....B
Leaden Flycatcher V
Magpie-lark I
LHI Grey Fantail X
Willie Wagtail V
Black-faced Cuckoo-shrike V
White-winged Triller V
Olive-backed Oriole V
Masked Woodswallow V
White-browed Woodswallow ...V
LHI Pied Currawong ....B
Australian Raven V
Skylark V
Richard's Pipit V
Common Chaffinch V
European Greenfinch V
European Goldfinch V
Common Redpoll V
Yellowhammer V
Welcome Swallow ....B
Tree Martin V
Fairy Martin V
LHI White-eye ....B
Robust White-eye X
Common Blackbird ....B
Island Thrush I
Song Thrush ....B
Tasman Starling X
Common Starling ....B

Index

Australian Kestrel 20
Australian Magpie Lark 35
Bar-tailed Godwit 48
Black-browed Albatross 38
Black Noddy 16
Black-winged Petrel 7
Blackbird 32
Breeding table seabirds 4
Brown Noddy 15
Buff-banded Rail 22
Cattle Egret 41
Double-banded Plover 43
Eastern Curlew 45
Emerald Ground Dove 26
Flesh-footed Shearwater . . .8
Great Cormorant 51
Grey Ternlet 18
Grey-tailed Tattler 47
Japanese Snipe 49
Kermadec Petrel 6
LHI Currawong 37
LHI Golden Whistler 34
LHI White eye 33
Little Shearwater 10

Map .28
Masked Booby 12
Masked Lapwing 23
Masked Owl 25
Pacific Golden Plover 40
Providence Petrel 5
Purple Swamphen 24
Red-necked Stint 50
Red-tailed Tropicbird 13
Ruddy Turnstone 42
Sacred Kingfisher 30
Song Thrush 31
Sooty Tern 14
Starling 36
Wandering Albatross 39
Wandering Tattler 46
Wedge-tailed Shearwater . . .9
Welcome Swallow 27
Whimbrel 44
White Tern 17
White-bellied Storm Petrel .11
White-faced Heron 19
Woodhen 21

KEY
B . . . . breeding species
R . . . . . . regular visitor
IR . . . . . . Irregular visitor
V . . . . Vagrant

I . . . . Introduced
X . . . . extinct
IX . . . . . . Introduced extinct

The Author

Photo courtesy Brendan Read

Ian Hutton has been observing and photographing the birds of Lord Howe Island since 1980. He has been location consultant for a number of television documentaries concerned with the birds of the Island, and has also been involved in a number of scientific projects with seabirds on the Island.

In 1982 he started nature walks on the Island and you can join one of these while visiting Lord Howe Island; or join one of several 8-day tours he runs each year for bird enthusiasts.

Write to:
Lord Howe Island Nature Tours
PO Box 157
Lord Howe Island
NSW 2898